TOP-SECRET NATURE

HOW SEEDS SPROUT

MARIE ROGERS

PowerKiDS press.

New York

Published in 2021 by The Rosen Publishing Group, Inc.
29 East 21st Street, New York, NY 10010

First Edition

Editor: Amanda Vink
Book Design: Rachel Rising

Portions of this work were originally authored by Evan Rhodes and published as *How Do Seeds Sprout?* All new material in this edition authored by Marie Rogers.

Photo Credits: Cover, p. 1 lovelyday12/Shutterstock.com; pp. 4,6,8,10,12,14,16,18,20,22 (background) cluckva/Shutterstock.com; p.5 WeStudio/Shutterstock.com; pp. 7,11,13 amenic181/Shutterstock.com; p. 9 Designua/Shutterstock.com; p.15 249 Anurak/Shutterstock.com; p. 17 HK-PHOTOGRAPHY/Shutterstock.com; p. 19 showcake/Shutterstock.com; p. 21 Bogdan Sonjachnyj/Shutterstock.com; p. 22 LiliGraphie/Shutterstock.com.

Library of Congress Cataloging-in-Publication Data

Names: Rogers, Marie, 1990- author.
Title: How seeds sprout / Marie Rogers.
Description: New York : PowerKids Press, [2021] | Series: Top-secret nature | Includes index. Identifiers: LCCN 2019044726 | ISBN 9781725317499 (library binding) | ISBN 9781725317475 (paperback) | ISBN 9781725317482 | ISBN 9781725317505 (ebook)
Subjects: LCSH: Germination–Juvenile literature. | Sprouts–Juvenile literature. | Seeds–Juvenile literature.
Classification: LCC QK740 .R64 2021 | DDC 571.8/62–dc23
LC record available at https://lccn.loc.gov/2019044726

Manufactured in the United States of America

Some of the images in this book illustrate individuals who are models. The depictions do not imply actual situations or events.

CPSIA Compliance Information: Batch #CSPK20. For Further Information contact Rosen Publishing, New York, New York at 1-800-237-9932.

CONTENTS

A World of Plants

Plants are an important part of our world. Without plants, animals wouldn't be able to live. Even though plants look different, most start out as seeds. What is a seed and how does it grow?

Seeds

Fully grown plants create seeds. These seeds spread to different places with help from animals and the wind. We can get seeds from plants or we can buy them at the store. If we plant them in the ground or in a pot, they can grow!

seed

The Seed Coat

All seeds have a seed coat, a hard shell that guards the insides. Within the seed, there's everything a baby plant needs to grow. The seed coat keeps the baby plant safe from dangers such as bugs and the weather.

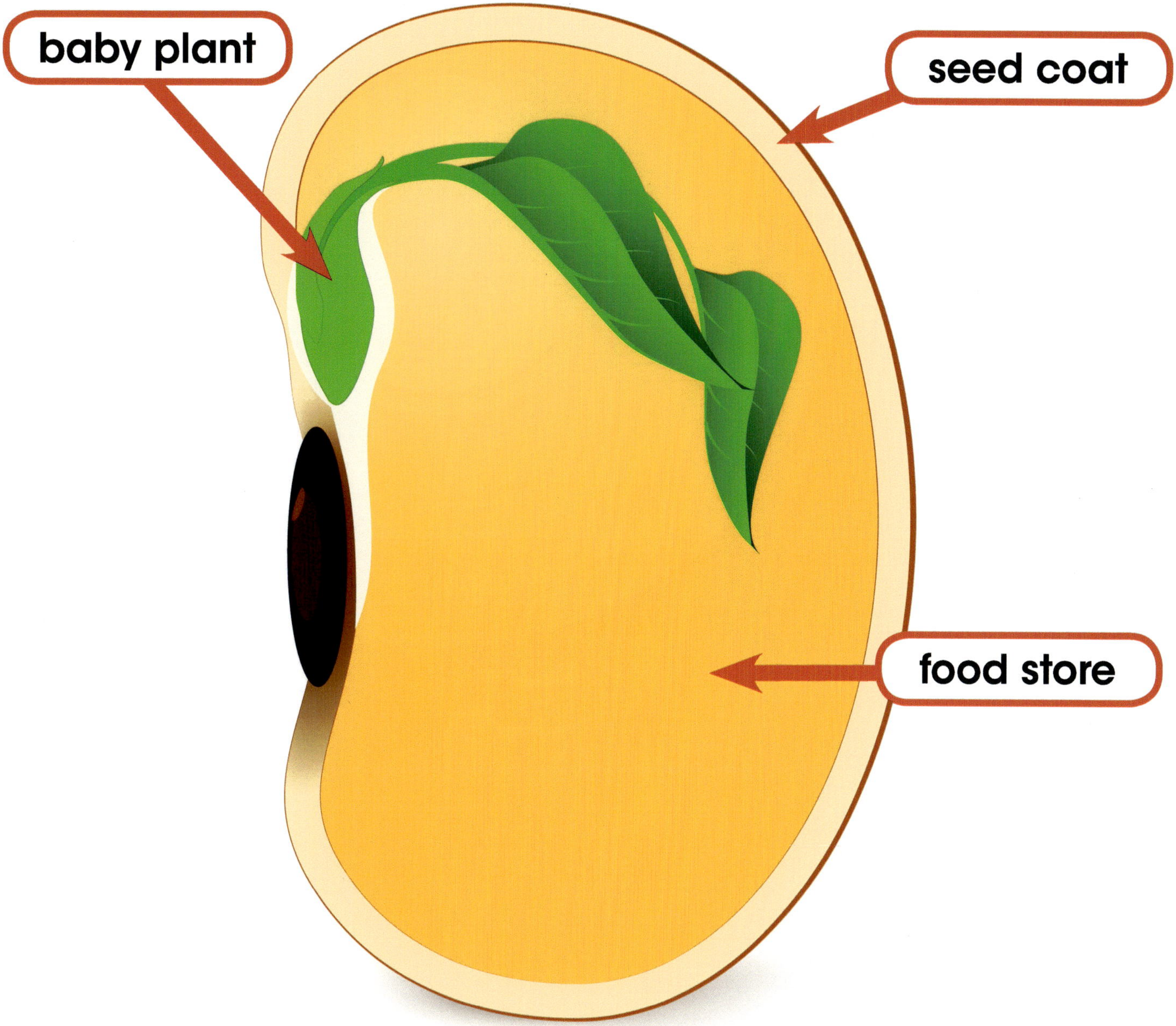
baby plant
seed coat
food store

Little Sprout

Seeds also have food the baby plant uses until it's ready to sprout. Sprouting is when something starts to grow. It happens a few days after the seed was planted. Scientists call this **germination**.

sprout

A Seed Needs

Seeds need three things in order to grow: water, air, and warmth. Some seeds need light to sprout. Other seeds need to be planted in the ground, where the dirt gives the seed food.

Growing

Before they start to grow, seeds need their surroundings to be warm. That's why most plants grow in the spring and summer. When it's cold, seeds stay **dormant**. They need to wait for better **conditions** and warmth before they begin to grow.

Water

A few days after someone plants a seed, it begins to sprout. This often happens under the ground, so we can't see it yet. The seed needs water, which makes the seed soft and helps open the seed coat. Now the seed begins to wake up!

Roots

The **seedling** first grows roots that reach deep into the dirt. These roots help the baby plant bring in water and hold itself in place. Depending on the type of plant, the seedling may have one or many roots.

roots

The Stem

Next, the plant's stem grows toward the light. In a few days, it will grow above the ground. The stem helps the plant stand up. The stem also **delivers** water and food to the leaves and flowers.

Adult Plants

Once the plant is above the ground, it keeps growing! It becomes an adult plant. Leaves collect sunlight to make food. The adult plant makes seeds, which become new plants. You can learn a lot by watching plants grow.

GLOSSARY

condition: The way things are at a time or in a place.

deliver: To take something to a certain place.

dormant: Not active for the time being.

germination: The act of a seed starting to sprout or grow.

seedling: A young plant grown from a seed.

INDEX

WEBSITES

Due to the changing nature of Internet links, PowerKids Press has developed an online list of websites related to the subject of this book. This site is updated regularly. Please use this link to access the list: www.powerkidslinks.com/tsn/seeds